Environmental Division 2024

Held at the 2024 AIChE Spring Meeting and 20th Global Congress on Process Safety

New Orleans, Louisiana, USA
24-28 March 2024

ISBN: 979-8-3313-1704-1

Printed from e-media with permission by:

Curran Associates, Inc.
57 Morehouse Lane
Red Hook, NY 12571

Some format issues inherent in the e-media version may also appear in this print version.

Printed with permission by Curran Associates, Inc. (2025)

For permission requests, please contact AIChE
at the address below.

AIChE
120 Wall Street, FL 23
New York, NY 10005-4020

Phone: (800) 242-4363
Fax: (203) 775-5177

www.aiche.org

Additional copies of this publication are available from:

Curran Associates, Inc.
57 Morehouse Lane
Red Hook, NY 12571 USA
Phone: 845-758-0400
Fax: 845-758-2633
Email: curran@proceedings.com
Web: www.proceedings.com

TABLE OF CONTENTS

CARBON MANAGEMENT AND SUSTAINABILITY

82a Lactic Acid Hydrophobic Natural Based Deep Eutectic Solvents for Carbon Capture at Wide Process Conditions 1
Ahmad Al-Bodour, Noor Alomari, Santiago Aparicio, Mert Atilhan

82b Unveiling the Truth: Energy-Environment-Economic (^{3}E) Assessment on Sustainability of Bioplastics and Recycling 2
Raffel Dharma Patria, Shao-Yuan Leu

82c Development of a Carbon Capture Membrane from an Existing Product and It's Laboratory Performance Under Highly Realistic Conditions 4
Christine Parrish, Joshua Watson, Max Roberts

ENVIRONMENTAL ISSUES AND CONTROLS IN SELECT INDUSTRIAL SECTORS I

102a Synthesizing Lithium Sorbents Using Coal Mine Drainage Solids 5
Chin-Min Cheng, Mengling Y. Stuckman, Christina Lopano, Maximilian Barczok, Wei Xiong, Bret H. Howard, Colleen Hoffman

102b Multi-Functional Sorbent Technology (MUST) for the Recovery/Removal of Critical/Heavy Metals from Fossil-Related Wastewaters 6
Walter Wilfong, Qiuming Wang, Fan Shi, Bret H. Howard, McMahan Gray

102c Techno-Economic Analysis of Recycling Rare Earth Metals from EV Motors 7
Maoqi Feng

ENVIRONMENTAL ISSUES AND CONTROLS IN SELECT INDUSTRIAL SECTORS II

122a Efficient Removal of Phenolic Compounds and Contaminants from Water Using Deep Eutectic Solvents-Based Adsorption Techniques 8
Neha Sawant, Abigail Maletta, Mert Atilhan

122b The Feasibility of Material Recovery from Processing the Entire Municipal Waste Stream 9
Tommy John

122c Engineering *C. Glutamicum* to Improve b-Ketoadipate Pathway Productivity 17
Tyler Okane, Arren Liu, Arul Mozhy Varman

ENVIRONMENTAL DIVISION POSTER SESSION

54a Environmental Assessment of Polycyclic Aromatic Hydrocarbons (PAHs) in the Polluted Coastal Cities 18
Ammar Bakheet

54b Geochemical and Spatial Distribution of Minerals in the Tailings of Ife/Ijesa Gold Mine Zone, Nigeria 19
Abiodun Fatoye, Albert Adebayo, Wasiu Tomori, Sunday Oladejo

54c Integra 2.0: Bridging the Gap from Global-Scale Contamination to Tissue-Specific Dose Assessment, Integration of a Computational Systems Biology Approach 20
Spyros Karakitsios, Achilleas Karakoltzidis, Fotini Nikiforou, Anna Aggaliadou, George Dimitriadis, Alberto Gotti, Dimosthenis Sarigiannis

54d Análisis De Escenario Natech: Derrame De Hidrocarburo Desencadenado Por EL Impacto De Fenómeno Hidrológico Extremo EN Tanque De Almacenamiento 23
Julio Alfonso DEL Vasto Urueña Sr., Carolina Lopez Perez

Author Index

82a Lactic Acid Hydrophobic Natural Based Deep Eutectic Solvents for Carbon Capture at Wide Process Conditions

Ahmad Al-Bodour, Noor Alomari, Santiago Aparicio, Mert Atilhan

Carbon dioxide (CO_2) solubility and the kinetics of absorption of CO_2 by Natural Deep Eutectic Solvents (NADES) were explored in this work. Three systems were under focus namely, carvone-lactic acid (1:1), cineole-lactic acid (1:1), and cineole-lactic acid (3:1). The CO_2 absorption conditions in the course of this work were a pressure range of 4.5 to 37 bar and at isotherms of 25 and 35 °C. The quantification of the CO_2 was in terms of mole fraction of the gas in the NADES. The process demonstrated that the absorption is influenced by the pressure positively and negatively by the temperature. Among the studied NADES systems, cineole-lactic acid (3:1) exhibits the highest capacity of CO_2 absorption at all the temperatures. A comparison was carried out between the experimental data and the COSMO-RS predictions. That revealed a systematic deviation in which all the experimental results were higher than the predictions and that deviation increases at the lower temperatures and at the higher pressure values. The kinetics of absorption also was investigated which showed that increasing the pressure made the absorption pace faster. Moreover, the system of cineole-lactic acid (1:1) showed faster absorption at 35 °C than that at 25 °C, which can be justified by the viscosity reduction. Also, heat of absorption for the tested systems was calculated based on Clausius-Clapeyron equation. The values of the heat of absorption were in the physi-sorption range and lower than the values of monoethanolamine (MEA). Heat of absorption magnitudes indicate the lower energy-intensive process of regeneration for the produced NADES. Generally, the findings direct to these NADES ability to capture CO_2 with reduced environmental effect and effective energy exploitation.

82b Unveiling the Truth: Energy-Environment-Economic (^{3}E) Assessment on Sustainability of Bioplastics and Recycling

Raffel Dharma Patria, Shao-Yuan Leu

Are bioplastics sustainable? The empirical evidence surrounding this debate abounds, but the answer is inconclusive. Comparability among the studies has been the issue.

This paper presents an unprecedented energy, environment, and economic (^{3}E) assessment of 16 studies on the life cycle and techno-economic assessments of bioplastics. All end-of-life scenarios were evaluated including landfill, incineration, composting, mechanical recycling, and chemical recycling. For benchmark, ^{3}E factors of petroleum-based plastics were compiled. The research answers the question by concentrating on three factors, namely non-renewable energy use in MJ/kg-plastic (energy), global warming potential in kg-CO_2e/kg-plastic (environment), and profitability in US$/kg-plastic (economic).

It was revealed that comparable outcomes can only be achieved when careful categorization is applied to sensitive elements such as database, methodology, biogenic carbon uptake, methane release during biodegradation, production scale, and capability of recycled plastic to substitute virgin plastic. From these revelations, we proposed a novel standardization method in this study. Sixteen existing studies in literature were carefully selected based on similar database and methodology of life cycle and techno-economic analyses. Then, we performed regression and cost-benefit analyses to reliably identify new trends and compare ^{3}E factors of different processes.

Following our proposed novel standardization protocol, the ^{3}E assessment demonstrated that bioplastics are sustainable only under specific conditions. When biogenic carbon is accounted for and no methane is released during biodegradation, the life cycle of bioplastics showed significantly lower global warming potential. Lower non-renewable energy use was found in production of bio-based plastics with average value of 39 MJ/kg-plastic, while average value for petroleum-based plastics is 80.2 MJ/kg-plastic. Regression analysis in this study revealed that inclusion of biogenic carbon uptake reduces the global warming potential of bioplastics by 2.1 kg-CO_2e/kg-plastic, and a linear correlation can be observed where global warming potential increases linearly with non-renewable energy use. In terms of economic performance, as the average costs of lignocellulosic biomass (US$0.8/kg) is lower than polyethylene (US$1.7/kg), it is profitable to produce lignocellulosic-based plastic with strong polymeric linkages without major alteration.

Recycling was deemed the most sustainable plastic disposal method in all three factors. Compared to direct energy recovery (incineration), recycling of 100,000 ton-plastic/year resulted in reduction of non-renewable energy use by 51.0-63.0 MJ/kg-plastic, reduction of global warming potential by 3.2-3.8 kg-CO_2e/kg-plastic and increase in profitability by US$237.0-330.3/kg-plastic. It was also found that recycling was highly affected by production scale. When production scale decreases from 100,000 to 40,000 tons-plastic/year, the profitability decreases from 403.6 to 75.9 US$/ton-plastic. However, it is important to note that most existing studies in literature assumed that recycled plastics can substitute virgin plastic. In reality, mechanically recycled plastics often have lower quality than virgin plastics. Furthermore, applications of recycling are currently hindered by technological limitations. Automatic sorting and efficient logistics for waste collection are necessary to achieve continuous and sustainable recycling.

It is hoped that the standardization method developed in this study not only enables reliable ^{3}E comparison of existing literature data, but also provides a guideline for future assessments on bioplastics. However, this method has a limitation as it only considers the ^{3}E factors. It is suggested that future research consider factors such as eutrophication and land use requirements for bioplastics to obtain a more comprehensive sustainability evaluation.

82c Development of a Carbon Capture Membrane from an Existing Product and It's Laboratory Performance Under Highly Realistic Conditions

Christine Parrish, Joshua Watson, Max Roberts

Compact Membrane Systems (CMS) has developed a novel Optiperm™ membrane platform to enable the efficient separation of two of the most critical needs of the industry: olefin-paraffin separation and carbon capture from post combustion flue gas. Carbon capture and storage (CCS) is considered to be a highly promising technique to reduce CO_2 emissions from point sources across multiple industries where membranes are seen as one of the most promising, low cost emerging technologies to compete with traditional amine towers.

CMS was able to build upon the success of the Optiperm™ olefins platform to create a new membrane technology capable of removing CO_2 from low pressure flue streams with high efficiency, creating a solution for point source carbon capture that required both low CAPEX and OPEX. Optiperm™ carbon applied the learnings of facilitated transport and humidification and transferred that for the selective removal of CO_2. In order to build a new membrane product offering off an existing product, modifications were needed to reach the performance necessary to achieve low cost carbon capture separation. We will discuss the effects of these modifications, such as moving from a spiral wound to a hollow fiber module, the effect of pressure drop, and overall system design to optimally run under very low driving forces.

CMS will present the results of laboratory testing for the separation of CO_2 from N_2 under highly realistic conditions. Specifically, we will discuss the effects of high feed temperature and very low feed pressure. We will also discuss the importance and effects of high stage cuts for facilitated transport membranes and the impact it has on the overall membrane performance. Finally, membrane stability data with common flue gas contaminants such as SO_x and NO_x will also be presented. Preliminary results from three pilot projects in steel, refractory materials, and acid gas incinerator applications will also be presented and compared to laboratory results.

102a Synthesizing Lithium Sorbents Using Coal Mine Drainage Solids

Chin-Min Cheng, Mengling Y. Stuckman, Christina Lopano, Maximilian Barczok, Wei Xiong, Bret H. Howard, Colleen Hoffman

Lithium (Li) is a critical mineral essential in many of emerging clean energy technologies. Direct extraction of Li from geothermal brines and oil/gas produced water is a distinctively more sustainable approach than hardrock leaching and evaporation pond extraction with respect to land, water, and energy uses. Inorganic molecular sieve sorbent, such as lithium manganese oxide (LMO) and lithium aluminum layered double hydroxide chloride (Li/Al LDH), is currently the most technologically advanced method for direct lithium extraction. Hydrogen manganese oxide (HMO), an ion sieve derived from LMO, has shown superior capacity and selectivity towards lithium. In this study, we prepared HMO from coal mine drainage (CMD) solids and examined the practicability of recovering Li from oil and gas produced waters and geothermal brines. We aim to innovate high-value added products from the existing extract REEs and CMs processing chain of legacy byproducts associated with coal production. CMD solids, which are precipitates collected from exhausted limestone gravels of passive CMD treatment beds, are enriched in rare earth elements (REEs) and other critical minerals (CMs), and therefore, considered as a viable alternative source for REEs/CMs to supply the Nation's demands. A patented extraction process developed by NETL (TREE®) has been demonstrated as a cost-effective and efficient approach to recover REEs/CMs from CMD solids. Here, we selected residuals of CMD solids from the TREE® process that are enriched in manganese (Mn) (>20% w/w) and successfully prepared LMO using a solid-state method modified from Yoshizuka et al. (2002). HMO sorbent was then obtained from the LMO precursor after replacing Li^+ with H^+ in a 0.1M hydrochloric acid (HCl) solution. Formation of spinel manganese oxide in the CMD-derived HMO sorbents was identified by X-ray diffraction analysis and Raman spectroscopy. By characterizing the adsorption of Li at three pH levels, i.e., 2.8±0.3, 7.0±0.7, and 9.7±0.5, we demonstrated the maximum adsorption capacity of a selected CMD-derived HMO sorbent, determined by fitting the Langmuir isotherm model, is similar to that of an HMO sorbent prepared from pure chemical reagents across a wide range of pH conditions. We demonstrate it is technically feasible to beneficially use the high-volume/low-value mineral residuals from REE/CM extraction to produce highly selective HMO sorbent for direct Li extraction. We continue to modify LMO synthesis conditions, e.g., Li-to-Mn ratio, calcination temperature and duration, and CMD solid feedstocks, to maximize the adsorption capacity of the CMD-derived HMO sorbent, whose adsorption selectivity in a solution with more complex constituent matrix, i.e., synthetic and actual geothermal brine and produced water, will be further characterized.

102b Multi-Functional Sorbent Technology (MUST) for the Recovery/Removal of Critical/Heavy Metals from Fossil-Related Wastewaters

Walter Wilfong, Qiuming Wang, Fan Shi, Bret H. Howard, McMahan Gray

Ever-increasing contamination of Earth's precious water systems with heavy and critical metals brings both a tremendous environmental burden to eliminate the toxic species and a golden opportunity to recover the most valuable elements. Fossil-related wastewaters like acid mine drainage (AMD), effluents from hydraulic fracturing processes (produced water), and flue gas desulfurization (FGD) wastewater from coal-fired powerplants contribute significantly to stream and groundwater pollution. A collection of remediation and recovery technologies like those predicated on photocatalytic-, electric-, chemical-, membrane-, and adsorptive-based processes currently exists in different stages of development or deployment between these methods of water management. Among these technologies, solid sorbent processes embody an optimal balance between cost-effectiveness, environmental friendliness, and technological maturity. The U.S. Department of Energy's National Energy Technology Laboratory developed a suit of cross-linked, functionalized silica sorbents tailored to eliminate the most toxic metals regulated by the U.S. Environmental Protection Agency (EPA); critical metals identified by the U.S. Geological Survey (USGS); and harmful organics like the infamous perfluoroalkyl and polyfluoroalkyl substances (PFAS). An array of laboratory and field tests proved that the MUST sorbents removed ppb-level lead from drinking water, selenium from FGD wastewater below EPA limits, aqueous dyes and PFAS; and fractionated low ppm-level critical metals (CM) from AMD and simulated produced water. Proof-of-concept for commercial CM recovery was verified through obtaining milligram-quantities of purified Al solids from a sorbent multi-bed AMD field-site test. Commercial efficacy was further supported by achieving purified fractions of adsorbed Mn upon treating both the AMD and synthetic produced water with the laboratory multi-bed test unit.

102c Techno-Economic Analysis of Recycling Rare Earth Metals from EV Motors

Maoqi Feng

We are investigating the technical and commercial feasibility of the recycling of rare earth metals from post consumer products. Lab-scale experiments were previously conducted by Polykala Technologies LLC to establish the processing steps and design basis mass balance cases for different potential feeds. The current assessment focuses on one specific feed: recycling Neodymium (Nd) metal from NdFeB magnets taken from recycled electric motors.

Expected gross revenue for each 1,500 kg batch of recycled NdFeB magnets was estimated as $21,800, which comes almost completely from the value of the recovered Nd metal. Operating expenses were estimated to be least $15,400 per batch, for maximum net revenue of $6,400 per batch. Assuming 230 batches are processed per year, the maximum net revenue is $1,466,000 per year.

Capital expenses were estimated to be $5.32 MM. This total value was generated by estimating the cost of 60% of the pieces of individual equipment in the unit - including what is expected to be the most expensive equipment – then scaling that cost up to estimate the cost for the entire process. The capital expense required for 60% of the pieces of process equipment is $3.14 MM. Scaling this estimate up from 60% to 100% of the pieces of process equipment increases the estimated capital expense from $3.14 MM to $5.32 MM.

Acknowledgement

This project was supported by DOE SBIR Project Contract Number DE-SC0021544.

122a *Efficient Removal of Phenolic Compounds and Contaminants from Water Using Deep Eutectic Solvents-Based Adsorption Techniques*

Neha Sawant, Abigail Maletta, Mert Atilhan

Phenolic compounds, which are extensively present in industrial effluents and wastewater, pose significant environmental threats due to their toxicity and severely threaten aquatic ecosystems and human health. In this study, a novel approach employing deep eutectic solvents (DES) based adsorbents was developed to effectively remove phenolic compounds from water. Tetra n octylammonium bromide-Linoleic based adsorbent was synthesized using eco-friendly and biodegradable components, their composition was optimized to enhance the extraction efficiency. Extraction efficiency was compared using a series of DES formulations, the adsorbent was synthesized based on DES to Ethyl Acrylate Ratio and UV cured for further analysis. The DES-based adsorbents demonstrated an extraordinary affinity for phenolic substances. The adsorption process was thoroughly investigated, taking into account DES concentration, reaction time, stir speed, and temperature factors. The results demonstrated high adsorption capacity making them promising candidates for toxic phenolic compound removal. Additionally, the synthesized DES formulations were characterized for their reusability, emphasizing their sustainable nature. This study provides valuable insights into the application of DES-based adsorbents as environmentally friendly and efficient sorbents for phenolic compound removal, contributing to the advancement of green technologies in water treatment.

The Feasibility of Material Recovery from Processing the Entire Municipal Solid Waste Stream

Tommy John, Tommy John Engineering, Bandera, TX

Abstract

Recycling, or material recovery, is the key to sustainability. Processing the full municipal solid waste (MSW) stream through material recovery facilities (MRF) is a new paradigm for waste management that will rely on chemical engineering for implementation. Depositing material in a landfill or combusting in a Waste to Energy (WTE) plant, as is a common practice for MSW, should be replaced with MRF. Sustainable material recovery from waste streams requires rigorous sorting and reliable, efficient processes for conversion of sorted streams into high value products. Plastic recycling is less developed than for other recyclables, but the Pyrolysis and Methanolysis processes extend the range of plastic that can be effectively recycled from waste streams. The remaining organic material following removal of all recyclables can be upcycled into valuable oxygenated chemicals.

A preliminary analysis indicates that a MRF system that can recover the highest value for recovered materials will produce revenue that is five to six times the revenue from typical landfill tipping fees, thus creating a sustainable revenue stream to pay off the large capital investment for the facility while maximizing environmental benefits.

Background

Sustainable waste management requires robust Material Recovery Facilities (MRF) to replace landfills and Waste to Energy (WTE) plants. In developing areas, MSW is often dumped directly into the environment (land or water), and should be replaced with material recovery. Landfills have historically been the lowest cost option for waste disposal, when suitable land is available. WTE is more expensive, but is often used in regions with a high population density and limited land. Both practices are difficult to permit and face considerable public opposition.

In addition to the obvious loss of valuable materials, landfills have other disadvantages. Landfills require a lot of land, and are often in a less accessible location requiring transportation of waste for a considerable distance. Although they can be constructed and operated to minimize environmental impact, the capture of landfill gas is limited to 70 to 80% by the need to limit air ingress to prevent landfill fires, resulting in unavoidable but substantial emissions of carbon dioxide, methane, and VOC. The landfill gas can be converted to natural gas or electricity, but the efficiency is low and carbon dioxide emissions are not abated.

WTE plants generate electricity using a steam cycle, but are limited by corrosion and fouling constraints to lower steam temperature and pressure, which, along with parasitic losses from emission control equipment, limits the net conversion efficiency to 20 to 30%. This is quite low compared to current combined cycle units that have an efficiency approaching 60%.

Although WTE produces a high value product, the recovery efficiency is low, and capital investment is high.

In the last few decades, a parallel system of waste disposal, dedicated to collecting, sorting, and recovering metals, paper, glass, and plastic (only the items that can be mechanically recycled) has evolved. It has to be subsidized and has proven to be only partially effective.

Processing the full MSW stream in a MRF is the only way to maximize recycling rates and improve public cooperation with reducing waste, including plastic, in the environment. Gaining acceptance for anti-litter campaigns is easier than for recycling, which is more complicated. Single stream MRF represents a new paradigm for MSW disposal that requires the application of processes well understood by chemical engineers, but largely unknown within the incumbent waste disposal organizations

It is paramount to have processes that effectively recover the high value products from the sorted materials. As with all energy and disposal facilities, a MRF based waste processing facility will require a large initial capital investment. In order to pay off the bonds or other financing of the investment, a high value for the recycled material must be collected in addition to a tipping fee. The value of the recycled material also aligns with environmental quality because, if a product has a high value, it is derived from a scarce resource and/or it requires a lot of energy to produce.

The higher value recyclable materials include metals, paper, and plastic; the latter for which the technology is least developed. Waste plastics represent a unique challenge because of the diverse composition, but an opportunity because of the high value of the sorted components. No single process will optimally recover plastic wastes, but multiple processes, including mechanical, chemolysis, and pyrolysis can be combined to optimize recovery.

Transfer Stations

Transfer Stations are currently used to optimize logistics for moving collected MSW to disposal sites. In urban areas they are used to reduce haul distance for collection trucks, since, as noted, landfills are typically located outside the urban center. Rural counties and small towns use them for collection for transfer to distant disposal sites, because a well-designed landfill is too expensive to serve a small population. More sophisticated MRF may require a larger scale for economy and location away from metro area justifying additional Transfer Stations integrated with MRF including rail or intermodal access.

Material Recovery Facilities

Digital technology enables improved sensors, controls, monitoring, artificial intelligence (AI), and mechanisms, including robotics, thus continuing to rapidly improve the sorting technology used in MRF. Rigorous sorting can direct components of the waste stream for optimal recovery as high value products. Dense solid materials such as glass, ceramics, and rocks have minimal commercial value, but are removed to improve downstream processing.

Metals, paper, and glass recycling are more advanced than plastic recovery. The processing of plastic is challenging because of the wide range of compositions; optimal recovery of plastic requires multiple processes.

Material Recovery

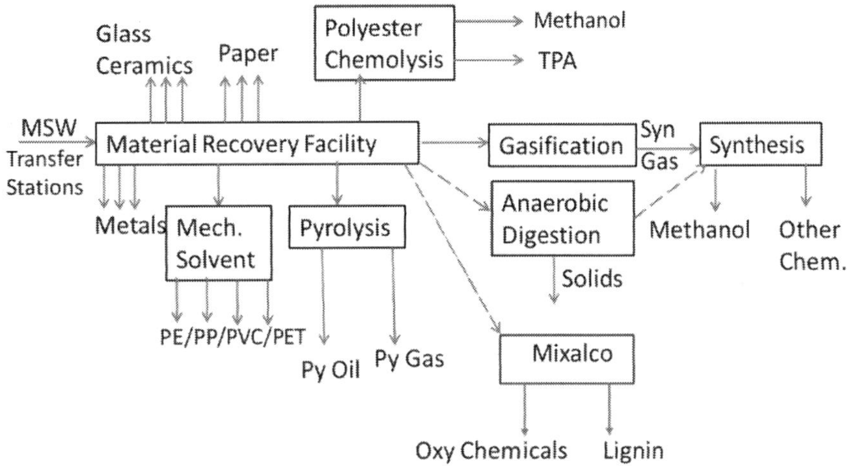

The remaining materials, which may be more than half of the original waste stream, are mostly biomass derived, or similar to biomass in composition: food waste, yard waste, wood, and uncaptured paper and plastic. The stream will have a high moisture content, which is not necessarily a disadvantage for the downstream processes; gasification, anaerobic digestion, or carboxylate salts, but the product yield will be based on the dry material content.

It should not be inferred that all of the elements in the schematic above are co-located with the MRF. The exceptions are the plastic pyrolysis and the remaining material recovery process because they are required to capture material that is not otherwise collected for recycle.

Waste Plastic Processing

Mechanical recycling, by which the material is ground to a powder and then extruded into pellets, is used to recover PE, PP, PET, PVC, and Polystyrene. It is currently the most widely used process, but is limited by the requirement of high purity, clean material. Pyrolysis of mixed hydrocarbon plastics is particularly useful because it effectively converts many of the polymers remaining, after diversion into other recycling processes, into Py oil, Py gas, and coke/asphalt, and tolerates contamination and mixtures. Chemolysis can convert Polyesters that are not suitable for mechanical recycling, including PET and fibers, into plastic feedstocks.

Pyrolysis

Pyrolysis of plastic waste will be an important part of the full material recovery process. This is not a new process; there has been a substantial amount of research, pilot plants, and even a few commercial plants implemented over several decades. The technology is currently receiving a lot of attention from major hydrocarbon processing operators and is available through multiple major process licensors. As of mid-2023, over 300,000 tonnes per year were operating, in Europe and North America, and 1.3 million tonnes per year of additional capacity was in development.

Pyrolysis processes extend the range of plastic that can be effectively recycled from waste streams (1). It performs best for hydrocarbon-based plastics, but it is effective with streams that have more diverse polymer compositions and contamination than can be tolerated with mechanical recycling. The main product is an oil stream with a composition that ranges from heavy oil to light naphtha that can be used as a refinery feedstock that is equivalent to petroleum. It can also be used as liquid feed to an olefins plant. An important secondary product is a gas stream that is rich in light olefins that can be injected into an olefin plant for complete recovery, if the location has pipeline access to such a facility. Other-wise the gas can be compressed and the C3+ components condensed for storage and shipment in pressure tanks.

Polystyrene that cannot be mechanically recycled can be converted into styrene monomer by pyrolysis when fed separately from other polymers.

Chemolysis

Eastman Chemicals is successfully operating a new process (Methanolysis) to convert polyesters (including PET) into mono-ethylene glycol (MEG) and dimethyl terephthalate (DMT), intermediates for production of new polymers.

PVC

PVC can be mechanically recycled, but advanced recycling is not economic because the polymer is 56% chlorine which cannot be recovered in a useful form. It can be downcycled into products such as synthetic construction materials which have commercial value.

The combination of these processes will recover almost the entire spectrum of plastic materials. Processes are being developed for other plastics.

Value of Recycled Plastic Products

The market value of the products produced by plastic recycling will depend on the uses. Pyrolysis complements rather than replaces Mechanical Recycling; the latter replaces polymer while the former is a feedstock. In the special case of polystyrene separated from the waste stream, it can be converted by pyrolysis into styrene monomer and used as feed to a polymerization process. PET can similarly be recycled using Chemolysis to produce the feed for polymerization.

In the special case of proximity to an olefins plant, the Py Gas can be fed to the compression section for full recovery of the stream. All of the constituents (alkanes, olefins, and hydrogen) are also in the cracked gas and will be recovered as product, feed, or fuel. Note that the pyrolysis plant does not have to be located within the olefins unit, but simply close enough for the gas to be transferred by a single pipe. The value of the py gas is the value of the individual constituents minus the cost of compression and processing. In the more common case of a remotely located Pyrolysis plant, the py gas can be compressed and cooled to partially condense a C3+ hydrocarbon stream containing primarily light olefins and alkanes that can be shipped in pressure vessels using ground transportation. In this case, ethylene, which is about 10% by mass of the py gas stream, would have only fuel value.

The use of Py Oil as a steam cracker liquid feed has been demonstrated. An alternative for any fraction that does not meet the specifications for steam cracker feed can be sent to a refinery. Compared to conventional crude oil, it is lighter, has more material in the range of transportation fuels, is low sulfur, and has lesser contaminants. Refineries have more options for processing and blending to meet fuel and feedstock specifications.

Processing of Remaining Organic Material

There are multiple options for processing the material remaining after the metals, glass, and almost all of the paper and plastic have been removed. It can be fed to an oxygen blown, high pressure gasification unit for conversion to syngas composed primarily of carbon monoxide and hydrogen that can be used to produce organic oxygenates. Methanol is an example, but it is a high-volume chemical produced with the economic advantage of economy of scale. Downstream oxygenated chemicals would have a higher inherent value because of additional processing and lower volume (less economy of scale for competitors).

There are other options for conversion of the unsorted materials, including anaerobic digestion, and fermentation into carboxylate salts (commonly known as the Mixalco process). The latter two bioprocesses will produce a residue of primarily lignin that would require further processing.

Revenue

The chart below shows the market value of key commodities produced in the MRF system based on pricing during February of 2024. The tipping fee (60 $/ton) is the US landfill average for 2022, up about 10% from 2021, after being relatively consistent for the preceding few years. The oil price is based on a recent value of $72 per barrel for West Texas Intermediate. The benchmark crude oil is slightly heavier and contains more sulfur than typical Py Oil, but the Py Oil is more olefinic. The oxygenated organic chemical price is based on recent US Gulf Coast pricing of Methanol as well as Ethylene Glycol and TPA. All other values are recently published prices for recycled commodities. Of course, all of these commodities will fluctuate in price.

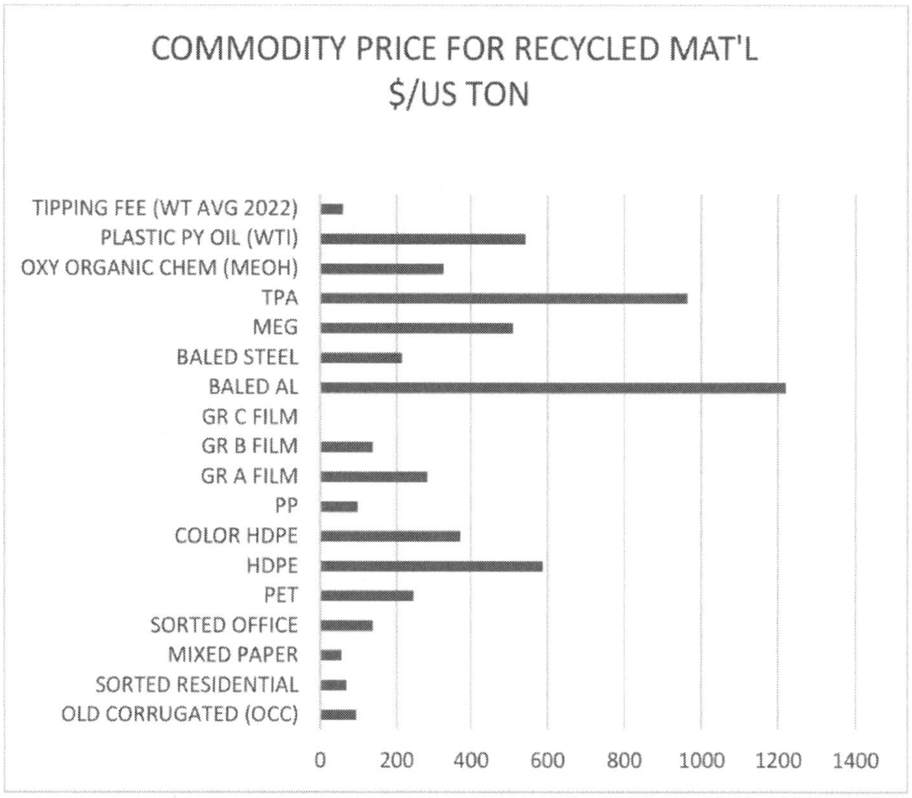

The following chart shows the economic contribution of each element of the MRF system based on a representative composition of the municipal waste stream, an assumed recovery percentage for each process, and the commodity prices shown in the preceding chart.

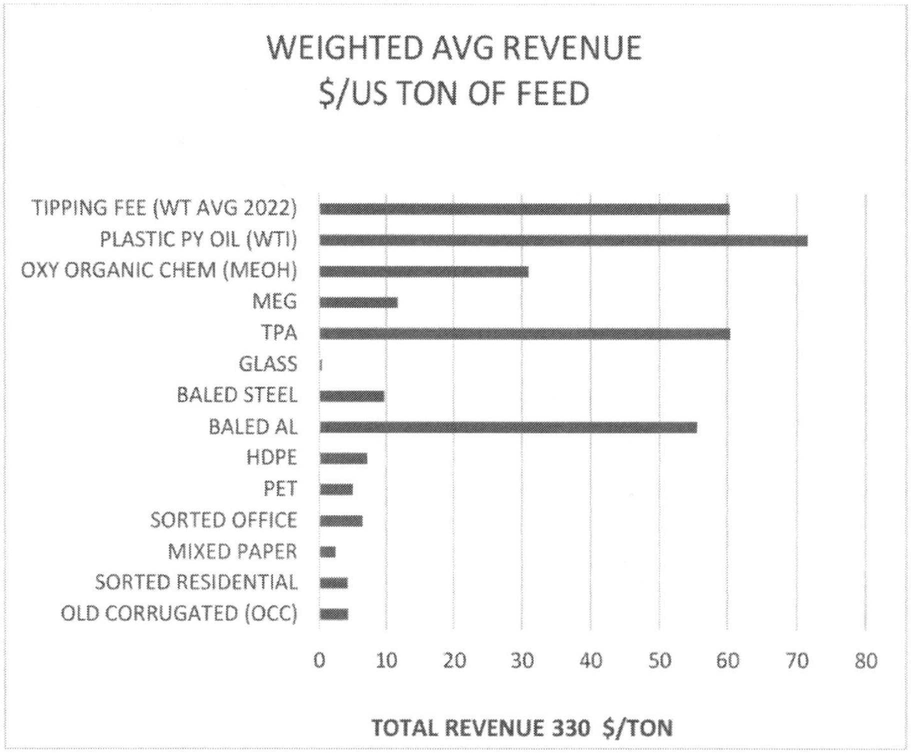

This indicates that the total revenue for the MRF is 5 to 6 times the current cost of disposal. Please note that the intent of this analysis is a preliminary, high-level view of the potential for full recovery of the MSW stream. While the figures may not be precise, the importance of capturing the highest value for each output is illustrated. And while the value of individual commodities will fluctuate, the total revenue estimated has varied by 10% or less over the past two years.

Investment

Following is an analysis of a facility for full recovery of a 3000 ton per day MSW stream, which would serve a population of several million people. A large facility is assumed to achieve economy of scale for the processes. The facility would include MRF, Pyrolysis, Methanolysis, and Gasification for Synthesis of Methanol. Sorted products including all metals, glass, and paper as well as plastics recoverable by mechanical recycling would be sold to offsite recycling facilities. The output of the Pyrolysis process is valued as crude oil; output of Methanolysis at the proportion of ethylene glycol and TPA produced, and output of gasification/synthesis as methanol. It is assumed that the cost of delivering the waste to the MRF would be equal or less than current cost for delivering to a landfill. The accuracy of the investment cost is challenging; the only public information is from a few project announcements. The analysis is obviously high-level early stage. Thus a 40% contingency for inflation and other uncertainties is assumed. Expenses are assumed to be 7% of total capital for Operations and Maintenance and 2% of capital for taxes and insurance. The Grand Total investment is large; almost a billion US dollars. A municipal waste project should be eligible for tax free bond financing.

	CAPACITY, US TONS/DAY	INVESTMENT, MILLION US $
MATERIAL RECOVERY FACIITY (MRF)	3000	180
PLASTIC PYROLYSIS	387	155
METHANOLYSIS	239	109
GASIFICATION & SYNTHESIS	449	216
METHANOL OUTPUT	235	
SUB-TOTAL		660
CONTINGENCY & INFLATION	40%	264
GRAND TOTAL		924
ANNUAL REVENUE		327
EXPENSES, O&M PLUS T&I	9%	(83)
PAYOUT, YEARS		3.8

Conclusions

- Sustainable waste management requires rigorous sorting & maximizing the value of recovered materials from the full MSW stream.
- The environmental and economic benefits of material recovery are aligned.
- Waste plastics are valuable components of the waste stream, but lag other recyclables in recovery applications
- Rapidly advancing technology for MRF & enhanced recovery of material justify the large investment required.
- Material Recovery from the full waste stream is the best way to reduce waste, including plastic, in the environment.

References

1. John, Tommy, "Pyrolysis: Filling the Gap in Recycling Waste Plastic", Paper 639375, AIChE Spring National Meeting, April, 2022.

122c Engineering *C. Glutamicum* to Improve b-Ketoadipate Pathway Productivity

Tyler Okane, Arren Liu, Arul Mozhy Varman

With over 50 million tons of lignin generated as a waste product annually, it is seen as the most abundant renewable feedstock for aromatic molecules [1]. Through chemical pretreatment, lignin can be broken down into two major aromatics, *p*-coumarate and *trans*-ferulate. Due to its resistance to the toxicity of aromatic molecules, *Corynebacterium glutamicum* is capable of metabolizing aromatics [2]. Additionally, there are several valuable intermediates within the aromatic catabolism pathways of *C. glutamicum* including beta-ketoadipate, a precursor for nylon-like polymers [3]. By utilizing the native pathway in *C. glutamicum*, along with genetic modifications for optimization, we propose that the biomanufacturing of beta-ketoadipate can be enhanced. This project will focus on metabolically engineering *C. glutamicum* to improve the productivity of the beta-ketoadipate pathway, with the ultimate goal of enhancing the microbial utilization of lignin-derived substrates.

The beta-ketoadipate pathway plays a pivotal role in the microbial metabolism of aromatic compounds in *C. glutamicum*, serving as a key route for the degradation of lignin-derived aromatic compounds [4]. This pathway is a complex metabolic route responsible for the conversion of aromatic compounds into central metabolites, making it an attractive target for metabolic engineering to enable sustainable bioproduction. Through previous research, we have found that the 4-hydroxybenzoate → protocatechuate step is a major bottleneck in this pathway. Researchers have found an alternative enzyme to *pobA*, *praI* [5], that can also convert 4-hydroxybenzoate into protocatechuate. In this project, I compare the catabolic efficiency of two different strains of *C. glutamicum*. One strain with *pobA*, one with praI. After creating the strains, I will feed them different substrates including *p*-coumarate, 4-hydroxybenzoate, and *trans*-ferulate, and look for other potential bottlenecks in the pathway to improve beta-ketoadipate production.

I have already cloned *praI* into a production strain, the next step in this project would be to clone and purify the *pobA*. Once we confirm that we have the correct gene, I will begin to construct a plasmid with *pobA*. PCR and gel electrophoresis would need to be used in order to verify that our constructed plasmids are correct. The next step would be to transform my plasmids into *Escherichia coli*. This step is necessary because transforming straight into *C. glutamicum* would have very low efficiency. After this, I will extract the recombinant plasmid and transform it a second time into a specialized strain of *E. coli* that will demethylate the plasmid. *C. glutamicum* will not accept methylated plasmids so transforming into a demethylation strain of *E. coli* first is required. Next, I will extract the demethylated plasmid and finally transform it into *C. glutamicum* to make a second production strain. Once I confirm that the *C. glutamicum* has accepted my plasmid, and they are correct, I can begin to test beta-ketoadipate production.

I will begin by creating cultures with each strain containing my plasmids and feeding them 4-hydroxybenzoate, sampling each culture daily. Once they stop growing I will run the samples I collected on a high-performance liquid chromatography machine. This will show me how much beta-ketoadipate is being produced, and will also allow me to identify if the protocachuate bottleneck was resolved. Pending the results, I would like to feed my cultures other substrates including *p*-coumarate, *trans*-ferulate, a mixture of different aromatic substrates, and then lignin-hydrolysate in order to show real-world application. In addition to this, I would like to do a 13C analysis on the pathways in both strains to see whether any of the intermediates are leaking into the central metabolism of *C. glutamicum*.

54a Environmental Assessment of Polycyclic Aromatic Hydrocarbons (PAHs) in the Polluted Coastal Cities

Ammar Bakheet

Polycyclic Aromatic Hydrocarbons (PAHs) is considered as one of the persistent organic pollutants (POPs) in the environment. PAHs are anthropogenic source to all marine environments (air, water, sediment and biota). They are available from burned out material such coal, wood, or even food as well as incomplete combustion from oil and gases industry. PAHs are characterized by high toxicity, hydrophobicity and lipophilicity. moreover, they are easy to transfer and bio-accumulated in the living organisms. PAHs are easy to transfer due to the semi-volatility character; that is why they are exchanges between different matrices (air, water, sediment).

54b Geochemical and Spatial Distribution of Minerals in the Tailings of Ife/Ijesa Gold Mine Zone, Nigeria

Abiodun Fatoye, Albert Adebayo, Wasiu Tomori, Sunday Oladejo

The main objective of this research is to identify the geochemical and mineralogical characteristics potential of unexplored tailings around gold deposit region using spatial statistics and map modeling. Some physicochemical parameters such as pH, redox potential, electrical conductivity, cation exchange capacity, total organic carbon, total organic matter, residual humidity, Cation exchange capacity and particle size were determine from both the mine drains and tailing samples using standard methods. The physico-chemical parameters of tailings ranges obtained were pH (6.0 – 7.3), Eh (−16 - 95 Mev), EC (49 - 156 μS/cm), RH (0.20-2.60%), CEC (3.64-6.45 cmol/kg), TOC (3.57-18.62%), TOM (6.15-22.93%). The geochemical oxide composition were identified using Proton Induced X-ray emission and the results indicated that $SiO2>Al2O3>Fe2O3>TiO2> K2O>MgO> CaO> Na2O> P2O5>MnO>Cr2O3>SrO>K2O>P2O5$.The major mineralogical components in the tailing samples were determined by a quantitative x-ray diffraction techniques using Rietveld method. Geostatistical relationships among the known points were determined using ArcGIS 10.2 software to interpolate minerals concentration with respect to the study area. The Rietveld method gave generally Quartz value as 73.73-92.76%, Ilmenite as 0.38-4.77%, Kaolinite group as 3.19-20.83%, Muscovite as 0.77-11.70% with trace of other minerals. The high percentage of quartz is an indication of a sandy environment with loose binding site.

54c Integra 2.0: Bridging the Gap from Global-Scale Contamination to Tissue-Specific Dose Assessment, Integration of a Computational Systems Biology Approach.

Spyros Karakitsios, Achilleas Karakoltzidis, Fotini Nikiforou, Anna Aggaliadou, George Dimitriadis, Alberto Gotti, Dimosthenis Sarigiannis

The assessment of aggregate exposure is a complex procedure that demands comprehensive data at every stage of the methodological pipeline. Addressing these requirements is the core focus of the INTEGRA platform (Sarigiannis et al., 2014; Sarigiannis et al., 2016) which encapsulates the life cycle of a compound within a modeling framework, consolidating all available information essential for assessing the source-to-dose continuum.

INTEGRA is an online integrative computational software that effectively navigates the calculation route from environmental release to internal dose, considering time-dependent processes and incorporating a Markov chain Monte Carlo (MCMC) probabilistic analysis framework. With this abstract, we are pleased to introduce the INTEGRA 2.0 platform, featuring significant enhancements. The updated version which is to be released in the upcoming days includes a detailed level four multimedia model, models for microenvironments assessment, an updated and more comprehensive version of the PBPK/TK model, integration with the Next Generation Systems Biology models that will be linked to toxicokinetic models by incorporating dose-response curves, and advanced 3D QSARs models to address data gaps.

The multimedia framework within INTEGRA is a sophisticated design that comprehensively considers a wide array of interactions occurring at significant distances, profoundly influencing the fate and transport of environmental and industrial chemicals. INTEGRA aligns itself with the rigorous guidelines set forth by the European Chemicals Agency ECHA (2012), which lays the foundation for a methodological pipeline ensuring the safety assessment of chemicals. What sets INTEGRA apart is its versatility, operating across various spatial scales, spanning from local to regional, continental, and even global domains. It adeptly manages the intricate exchange of chemicals across multiple environmental media, encompassing air, soil, water, sediment, and their subsequent transfer into food items such as crops, meat, milk, and fish. In the updated version, we've elevated the INTEGRA multimedia model to a level four by incorporating considerations for the time required for a chemical to circulate within environmental components. This enhancement yields several benefits, including heightened precision, advanced multiscale modeling capabilities, detailed cross-media mass transfer analysis, increased customization and predictability, and improved integration with other models through real-time monitoring. These improvements collectively contribute to the platform's enhanced capabilities for comprehensive environmental and human health assessments. The framework presented considers emissions, advection, diffusion, and degradation, principles as well. It is worth mentioning that a variety of plant tissues have been included in the model as well, increasing even more the complexity of the mechanistic model as well as its accuracy and precision.

The modeling of indoor microenvironments and the comprehensive assessment of individual exposure in close proximity, often referred to as near-field exposure, have been significantly enhanced through the improvement of the two-zone model by Sarigiannis et al. (2012) that was used. This innovative model offers a comprehensive understanding of indoor microenvironments by accounting for the partitioning of substances among gaseous, particulate, and settled dust phases (Sarigiannis et al., 2012). The key strength of this exposure model lies in its holistic approach, considering all potential means and routes of exposure. Moreover, it factors in a multitude of variables that can influence exposure levels, such as age and gender, by incorporating activity-based inhalation rates, dietary patterns, food item intake rates, the quantity of soil and dust ingested, and children's hand-to-mouth behavior. This approach ensures a more accurate and nuanced evaluation of exposure, reflecting the complex interplay of environmental and behavioral factors on individual health risks.

The PBPK model has undergone significant expansion, featuring additional compartments such as the intestine, spleen, feces, hair, and a generic blood-brain barrier mechanism with six sub-compartments including red blood cells, central plasma, cells, cerebrospinal fluid, intracellular, and extracellular fluid. This expansion has necessitated the update of our dynamic mass balance equations. Importantly, all these included organs are now equipped with specific metabolism mechanisms tailored to individual chemicals. Furthermore, a comprehensive system of equations has been developed to account for age-related metabolism alterations, adding a layer of detail and precision to the model. In addition, the generic PBPK model is not only geared with a Michaelis – Menten approach but also with the Brigs-Haldane equation.

In the updated kinetic model, the direction of metabolism has evolved from a linear pathway with three metabolites to a multi-dimensional framework with an infinite number of metabolites. Users now have the flexibility to either select from the pre-existing toxicokinetic models with predefined metabolite compositions or design their custom metabolism pathways, with the option to incorporate an infinite number of branches as needed. The source code is implemented in R computing environment, enabling the use of text processing techniques to generate an infinite set of differential mass balance equations, which can be seamlessly integrated with the parent compound. Furthermore, the model introduces a mechanism tailored for particularly lipophilic chemicals, which allows them to penetrate cell membranes, and includes protein binding mechanisms to account for a broader spectrum of chemical behaviors and interactions within biological systems.

For the parameterization of pharmacokinetic models, high-precision 3D QSARs (Quantitative Structure-Activity Relationships) have been developed using deep learning architectures such as TensorFlow and Keras. These models encompass numerous parameters, including Blood: Tissue partition coefficients, enzyme kinetics, plasma protein fraction bound, and more. These models are made accessible to the user, offering an alternative when data necessary for running the PBPK model is not readily available. As a result, to execute only the pharmacokinetic model in Tier 1, users need the chemical's structural information (InChI identifier or SMILES) and details about the quantity of chemical in the three exposure routes. This approach ensures flexibility and precision in pharmacokinetic modeling, even in cases where comprehensive data may be limited.

To apply this model at the molecular level, the integration of Next Generation Systems Biology (NGSB) models, as presented in the work by Karakoltzidis et al. (2023), has been seamlessly incorporated into the INTEGRA methodological pipeline. These NGSB models represent extensive mathematical models consisting of hundreds of equations that intricately describe the functioning of biological systems in depth and precision. The successful connection of PBPK outputs with NGSB models includes the integration of chemical-specific dose-response curves for multiple biomarkers (genes, enzymes, proteins, endogenous metabolites), ensuring a comprehensive approach. The collection of dose-response curves in the INTEGRA database is facilitated by employing Natural Language Processing (NLP) and token classification techniques. These curves help establish links between specific biomarkers and adverse outcomes, allowing for tailored assessments based on available data and user preferences. The utility of INTEGRA extends across a multitude of applications in industry, not only at the risk assessment level but also within the realm of toxicology. It is a valuable tool for evaluating occupational exposure and assessing the potential consequences of accidents, providing a versatile and comprehensive platform for various industrial risk assessment needs.

ECHA. (2012). *Guidance on information requirements and chemical safety assessment. Chapter R.16: Environmental Exposure Estimation Version 2.1.*

Karakoltzidis, A., Karakitsios, S., & Sarigiannis, D. (2023). MINER: A Text Mining Approach for the Introduction of Next-Generation Systems Biology Models. *Toxicology letters*, *384*. https://doi.org/10.1016/S0378-4274(23)00538-6

Sarigiannis, D., Gotti, A., Karakitsios, S., Kontoroupis, P., & Nikolaki, S. (2012). Intera Platform: A Tool for Mechanistic Risk Assessment of Indoor Air Pollutants. *Epidemiology*, *23*(5S). https://doi.org/10.1097/01.ede.0000417237.68516.29

Sarigiannis, D., Karakitsios, S., Antonakopoulou, M., & Gotti, A. (2012). Exposure analysis of accidental release of mercury from compact fluorescent lamps (CFLs). *Science of the Total Environment*, *435–436*(0), 306-315. https://doi.org/10.1016/j.scitotenv.2012.07.026

Sarigiannis, D., Karaktitsios, S., Gotti, A., Loizou, G., Cherrie, J., Smolders, R., De Brouwere, K., Galea, K., Jones, K., Handakas, E., Papadaki, K., & Sleeuwenhoek, A. (2014). Integra: From global scale contamination to tissue dose. Proceedings - 7th International Congress on Environmental Modelling and Software: Bold Visions for Environmental Modeling, iEMSs,

Sarigiannis, D. A., Karakitsios, S. P., Handakas, E., Simou, K., Solomou, E., & Gotti, A. (2016). Integrated exposure and risk characterization of bisphenol-A in Europe. *Food and Chemical Toxicology*, *98*, 134-147. https://doi.org/10.1016/j.fct.2016.10.017

54d Análisis De Escenario Natech: Derrame De Hidrocarburo Desencadenado Por EL Impacto De Fenómeno Hidrológico Extremo EN Tanque De Almacenamiento

Julio Alfonso DEL Vasto Urueña Sr., Carolina Lopez Perez

La creciente industrialización trae consigo un incremento en el almacenamiento, manejo y comercialización de sustancias peligrosas, al mismo tiempo que implica la ocupación de nuevas zonas para la localización de instalaciones, cuya ubicación a menudo es seleccionada sin considerar la exposición a amenazas naturales, y por ende, a las fuerzas externas que tienen el potencial de desencadenar un accidente tecnológico y liberar grandes cantidades de agentes nocivos para la salud pública y el ambiente. Técnicamente este tipo de eventos se conocen bajo el término de NaTech (accidentes tecnológicos desencadenados por eventos de origen natural - Natural disasters triggering TECHnological accidents), (Cruz et al., 2004; Showalter & Myers, 1994).

En general, el diseño y operación de equipos, procesos, estructuras e instalaciones industriales de alto riesgo, sigue un conjunto de normas y estándares internacionales de la industria, que en términos de la seguridad de procesos, son propuestos para soportar las fuerzas operacionales (temperatura, presión, caudal, flujos, etc.), sin considerar la acción de fuerzas externas (Cozzani et al., 2009), como por ejemplo, aquellas ejercidas por la velocidad del agua (fuerza hidrodinámica) y la profundidad de la lámina de agua (fuerza hidrostática) sobre las unidades de proceso, durante una inundación.

La fase y peligrosidad de la sustancia implicada en un evento NaTech, depende del proceso, equipo e instalación afectada por el fenómeno natural, y la severidad del suceso en términos de cantidad liberada, dispersión y receptores afectados, depende del tipo de fenómeno natural que desencadena la pérdida de contención accidental, siendo las inundaciones, tormentas eléctricas y sismos (Krausmann et al., 2008), los principales iniciadores de este tipo de incidentes. En ese orden de ideas, la infraestructura petrolera, como industria de alto riesgo (Yanting & Liyun, 2011), cuya localización o emplazamiento en el contexto mundial, cubre vastas zonas del territorio, está expuesta a diferentes amenazas naturales, siendo los tanques de almacenamiento de hidrocarburos, los equipos más afectados (V. Cozzani et al., 2010) y con mayor criticidad por los altos inventarios de petróleo que potencialmente serían derramados en un evento de este tipo.

Entre los principales daños o afectaciones que sufren este tipo de equipos debido a las inundaciones, según Campedel (2009), se encuentran daños estructurales (desplazamientos, impactos con objetos flotantes, colapso) y fallas eléctricas en la sensórica que automatiza los procesos asociados, mientras que en las consecuencias más graves están aquellas relacionadas con la dispersión y reacción química de la sustancia peligrosa en agua.

En este sentido, la ausencia de estas barreras de seguridad en los estándares de diseño, construcción y operación de equipos y procesos, constituyen una condición de debilitamiento de las instalaciones frente a amenazas naturales y por ende, abre la posibilidad a la ocurrencia de un evento tipo NaTech, cuyo efecto unitario o dominó puede conllevar a la pérdida de contención masiva de sustancias altamente peligrosas.

Investigaciones realizadas por Antonioni et al., 2009, Antonioni et al., 2015, respecto a la evaluación cuantitativa del riesgo (ECR) por derrames de petróleo en tanques de almacenamiento debido a inundaciones, proponen el análisis a partir de la caracterización de la amenaza natural y de la vulnerabilidad de los tanques, este último en función de modelos de fragilidad que determinan la probabilidad de daño en el equipo, mientras que para el fenómeno amenazante plantean el análisis del evento hidrológico, a partir de las fuerzas externas ejercidas por la profundidad de la lámina de agua y por la velocidad de la corriente, asociadas a un determinado periodo de retorno, semejante a lo evidenciado por V. Cozzani et al., 2010 y V. Cozzani et al., 2014, en cuanto a la relación hallada entre la profundidad de la inundación y la ocurrencia de eventos NaTech.

En ese orden de ideas, la selección del evento hidrológico, como punto de partida del análisis, dependerá de los criterios establecidos en la evaluación en cuanto a la magnitud y probabilidad de la inundación de interés, en consideración a que la ocurrencia de estos sucesos está determinada por un periodo de retorno (t_r) en años.

A partir de allí, empleando métodos de simulación hidráulica a superficie libre, se obtienen profundidades (h_w) y velocidades (v_w) de la lámina de inundación que afecta el equipo en análisis (tanque de almacenamiento), teniendo en cuenta que mediante estas dos variables se simplifican los efectos hidráulicos asociados a las fuerzas externas ejercidas por la inmersión lenta, oleaje de alta profundidad, oleaje de velocidad moderada, oleaje de alta velocidad y finalmente oleaje de alta velocidad con profundidad limitada (V. Cozzani et al., 2014).

Posteriormente es necesario determinar el tipo y el nivel de daño ocasionado en el equipo expuesto, debido a la inundación. Para ello diversos autores han propuesto modelos mecánicos de falla que permiten estimar con cierta aproximación la vulnerabilidad del equipo, entre ellos Salzano et al., 2003, G. Antonioni et al., 2009, G. Landucci et al., 2012 y G. Antonioni et al., 2015, cuyas contribuciones se representan mediante curvas de fragilidad, modelos mecánicos de falla bajo carga de inundación.

Por su parte Rojas, S., 2005, Guerrero, Y., 2007 y Zapata C. et al.,2002, han evaluado diversas aplicaciones computacionales para analizar predictivamente el comportamiento de derrames de hidrocarburo en ambientes acuáticos, principalmente en aguas continentales en Colombia, a fin de dimensionar las consecuencias potenciales, empleando diferentes software de modelado de tipo Lagrangiano que convergen en el cálculo de los procesos de desgaste o intemperización del derrame y de los efectos de la hidrodinámica en el transporte del derrame, siendo comunes la estimación de la advección, trayectoria, destino y las pérdidas por evaporación, dispersión mecánica, adherencia a orillas o frontera terrestre y la disolución, en modelos con diferentes capacidades y tipos de licencia.

De otro lado, en un esfuerzo por estimar el comportamiento de los incendios de hidrocarburo que ocurren durante un derrame, Imazu & Nishino (2018) proponen considerar dicho derrame como un conjunto de partículas circulares que flotan ardiendo, cuyas posiciones temporales dependen del campo de velocidades y trayectorias de la inundación y que su ignición inicial es impredecible. La zona de combustión y el flujo de calor radiante para cada una de las partículas incendiadas se estima de manera homogénea e independiente de la dirección del viento, desde el centro de la partícula de hidrocarburo y cesa cuando el espesor de cada partícula del derrame alcanza 1 mm.

Las capas de protección aplicables a los equipos y procesos expuestos a las fuerzas externas ejercidas por los fenómenos naturales, de acuerdo con el concepto propuesto por CCPS (2001), como barreras de seguridad proporcionan soluciones técnicas para evitar, prevenir o mitigar las pérdidas de contención dentro de un amplio espectro de alternativas que van desde las barreras pasivas hasta las activas.

Lecciones aprendidas de eventos NaTech ocurridos en distintas partes del mundo, recogidas por Misuri et. al (2019) y diversos autores, ponen de manifiesto fallas comunes en las barreras de seguridad asociadas a desabastecimiento de agua de los sistemas de protección contra incendios

(Steinberg & Cruz, 2004), sistemas de supresión de vapores a base de agua y espuma no disponibles (Girgin, 2011), sistemas de contención de derrames colmatados (eMars), almacenamiento de químicos de refrigeración no disponibles (Additives for Polymers, 2017), válvulas de accionamiento manual inundadas e inaccesibles (NTSB, 1996) y generadores de energía de respaldo completamente inundados (Labib and Harris, 2015), entre otros.

De acuerdo con Necci A., Krausmann E. (2021), algunos ejemplos para tratar el riesgo asociado a las amenazas naturales pueden ir desde barreras físicas cómo: construcción de diques, construcción de canales artificiales, estabilización de riberas, compactación y estabilización de suelos, reorganización de instalaciones industriales y una correcta localización de la planta, mientras que en términos de medidas procedimentales, la definición de roles y funciones, así como la lista de acciones y tiempos de actuación, contribuirán a la reducción del riesgo.

Ahora bien, cuando sea posible anticiparse al impacto del evento natural en la instalación, a través de pronósticos y predicciones, la implementación de medidas como los apagados de emergencia, el establecimiento de un nivel de llenado crítico (CFL) de los equipos para contrarrestar la fuerza de la inundación, las evacuaciones de personal no vital operacionalmente, además de la comunicación permanente con las autoridades locales y organismos de socorro, junto con el monitoreo de la evolución de la amenaza, son determinantes.

Es así como el presente tema, tiene como objetivo analizar la fuerza ejercida por un evento hidrológico extremo sobre una instalación industrial, específicamente sobre un tanque vertical de almacenamiento de hidrocarburo refinado que hace parte de una planta de gas húmedo localizada en la planicie inundable de un río, empleando distintas formulaciones matemáticas para determinar el modo en que falla el recipiente en función de la carga hidrodinámica e hidrostática ejercida por la inundación, para posteriormente, a partir del uso de modelos computacionales bidimensionales y del uso de tecnología LiDAR para generar un modelado 3D de la planta, analizar las consecuencias desencadenadas por la pérdida de contención del producto, en términos del comportamiento y trayectoria del derrame y de la radiación térmica liberada por el incendio de piscina que genera dicho derrame.

AUTHOR INDEX

Adebayo, Albert....19
Aggaliadou, Anna....20
Al-Bodour, Ahmad....1
Alomari, Noor....1
Aparicio, Santiago....1
Atilhan, Mert....1, 8
Bakheet, Ammar....18
Barczok, Maximilian....5
Cheng, Chin-Min....5
Dimitriadis, George....20
Fatoye, Abiodun....19
Feng, Maoqi....7
Gotti, Alberto....20
Gray, McMahan....6
Hoffman, Colleen....5
Howard, Bret H....5, 6
John, Tommy....9
Karakitsios, Spyros....20
Karakoltzidis, Achilleas....20
Leu, Shao-Yuan....2
Liu, Arren....17
Lopano, Christina....5
Maletta, Abigail....8
Nikiforou, Fotini....20
Okane, Tyler....17
Oladejo, Sunday....19
Parrish, Christine....4
Patria, Raffel Dharma....2
Perez, Carolina Lopez....23
Roberts, Max....4
Sarigiannis, Dimosthenis....20
Sawant, Neha....8
Shi, Fan....6
Stuckman, Mengling Y....5
Tomori, Wasiu....19
Varman, Arul Mozhy....17
Wang, Qiuming....6
Watson, Joshua....4
Wilfong, Walter....6
Xiong, Wei....5

AIChE
120 Wall Street, FL 23
New York, NY 10005-4020

ISBN 979-8-3313-1704-1
9 798331 317041

CURRAN ASSOCIATES INC.
proceedings.com